上海市科学技术委员会"科技创新行动计划"
中国珍稀物种科普丛书

野马的故事

叶晓青　张　萍　王　丽　著
王　紫　绘
万倩倩　译

上海科学技术出版社

图书在版编目（CIP）数据

野马的故事：汉英对照 / 叶晓青，张萍，王丽著；王紫绘；万倩倩译. -- 上海：上海科学技术出版社，2021.10
（中国珍稀物种科普丛书）
ISBN 978-7-5478-5490-7

Ⅰ.①野… Ⅱ.①叶…②张…③王…④王…⑤万… Ⅲ.①马-少儿读物-汉、英 Ⅳ.①Q959.843-49

中国版本图书馆CIP数据核字(2021)第205820号

扫码，观赏"中国珍稀物种"系列纪录片《普氏野马》

普氏野马是世界上仅存的野马。它们命运多舛，曾被宣布野外灭绝。历经百年，它们重返故乡，但它们能否找回野性，适应先祖曾经生活的地方并重拾辉煌？请跟随"中国珍稀物种"系列纪录片《普氏野马》去新疆卡拉麦里找寻答案吧。

中国珍稀物种科普丛书
野马的故事

叶晓青　张　萍　王　丽　著
王　紫　绘
万倩倩　译

上海世纪出版（集团）有限公司
上海科学技术出版社　出版、发行
（上海市闵行区号景路159弄A座10F–9F）
邮政编码 201101　www.sstp.cn
浙江新华印刷技术有限公司印刷
开本 889×1194　1/16　印张 4
字数：70千字
2021年10月第1版　2021年10月第1次印刷
ISBN 978-7-5478-5490-7/N·226
定价：50.00元

本书如有缺页、错装或坏损等严重质量问题，请向承印厂联系调换

导读

阿布是在"新疆卡拉麦里山有蹄类野生动物自然保护区"新诞生的小野马,血液里流淌着的野性让它对一切事物都充满着无畏和好奇。一次偷懒,让阿布离开了家族马群,即将到来的寒冬和未知的危险,让身心俱疲的阿布感到绝望。

本书分为上下两个部分。第一部分采用儿童喜闻乐见的绘本故事形式,在尊重科学事实的基础上,将充满趣味的故事与精美的绘画相结合,提升整体艺术表现力,给读者文字以外的另一个想象空间。第二部分采用问答的形式,以增进读者对该珍稀物种的科学认识,通俗易懂的语言配上精美的照片,有利于儿童阅读和理解。本书是一本兼具科学意趣和艺术感染力的科普读物。

目录

我的家在卡拉麦里	6
野马的秘密知多少	58
我的名片	58
为什么叫作普氏野马	59
普氏野马是怎么回到故乡的	60
普氏野马是世界上唯一的野马吗	60
普氏野马最大的本领是什么	61
普氏野马喜欢吃什么	62
普氏野马什么时候战斗力最强	63
普氏野马是卡拉麦里最强悍的动物吗	64
普氏野马和蒙古野驴是好朋友吗	64

我的家在卡拉麦里

一场秋雨过后,荒漠迎来了一次返青。"国王"带领着他的马群努力啃食着地面上的植物。马群中有三匹出生没多久的小马。眼看着寒冬逼近,"国王"非常担心,因为冬天是一年中最难熬的季节。

The desert turned green after an autumn rain. A group of Przewalski's horses were nibbling the plants on the ground, among whom were three colts. King was very worried about the coming winter, as it was the hardest time in a year.

"国王"是这群普氏野马的首领。像这样的家族在新疆卡拉麦里还有20多个。新疆卡拉麦里山有蹄类野生动物自然保护区是中国最大的有蹄类自然保护区之一，这里是人类为他们建立的家园。

King was the leader of this group. There were more than 20 families like them in Kalamaili. Kalamaili Mountain Nature Reserve, XinJiang is one of the largest reserve for ungulates in China.

阿布在家族里年龄最小，也最调皮，平常非常喜欢的就是和另外两个兄弟追逐打闹。"国王"对这个最小的孩子很是头痛。血液里的野性让阿布对荒漠充满了好奇和无畏，因为擅自离开马群而挨骂已经是家常便饭了。

Abu was the youngest and the naughtiest child in the family. He liked playing with two brothers very much. King had no idea how to train him. The wildness in the blood made Abu curious and fearless. Being scolded for leaving without permission was nothing new for him.

卡拉麦里为数不多的水源地，聚集着荒漠中的野生动物，普氏野马、蒙古野驴、鹅喉羚、狼、赤狐都是这里的常客。进入深秋，大多数的水源会结冰，喝水将变得越来越困难。

Przewalski's horses, Asiatic wild asses, goitered gazelles, wolves, red foxes and other wild animals visited the water sources in Kalamaili frequently. In late autumn, to find enough drinking water became more and more difficult due to freeze.

阿布每天要跟随大部队移动上百千米去寻找牧草和水源，长时间的迁徙让他渐渐体力不支，前腿都抬不起来了。每次他想偷个懒休息一会儿的时候，爸爸都会严厉地训斥，让他必须跟上队伍。阿布觉得很憋屈，真想自己一个人自由自在地玩耍。

Abu's family had to move hundreds of kilometers every day to look for grass and water. After a long time of migration, Abu felt too tired to even lift his front legs. Whenever he wanted to rest, his dad would severely blame him. Abu wished to play by himself freely so much!

阿布走得实在太累了，趁大家喝水的时候，他从爸爸眼皮底下溜走，偷偷跑到水塘后的一个山丘躲了起来。心想："嘿嘿，在这里你们就找不到我了，我可以好好睡一觉，等睡醒了再出发。"

Couldn't bear any more, Abu slipped away and ran to a hill behind the pool when others were drinking water. "Aha! Now you can't find me, I'll just take a nap first."

阿布睡饱了,睁开大眼睛。这时天色已经完全暗了下来,一条璀璨的银河悬挂在深邃的夜空中。顾不得欣赏这美极了的景色,阿布撒开腿就往水塘跑去,边跑边嘀咕:"糟糕!又要挨骂了,不过至少我美美地睡了一觉,也算值得。"

After a sweet sleep, Abu opened his big eyes. It had been completely dark, with a bright Milky Way hanging in the sky. Abu ran to the pool and muttered, "Oh, no! They must blame me again, but at least I have had a good sleep."

可是等他跑到水塘边,却傻眼了:水塘边冷冷清清的,哪里还有马群的影子。阿布绕着水塘跑了三四圈,鼻孔喘着粗气,朝天拼命地嘶吼,却没有任何回应。阿布不敢走远,他希望马群明天早晨会回到这个水塘,这样他就能和家人团聚了。

However, nobody was there when he reached the pool. Abu searched around several times, gasping and yelling desperately, but there was no response. He dared not go far, as he hoped his family would return tomorrow morning.

天渐渐亮了，远处隐约出现了马群的身影。阿布一下子来了精神！可是等马群走近，阿布的希望落空了：这是另一个迁徙过来的马群。马群里的小马好奇地打量着阿布问："你怎么一个人，没和爸爸妈妈一起走呢？"

The sun rose and some horses appeared in the distance. Is that my family? But to his great disappointment, that was another group of migrated horses. One colt looked at Abu curiously and asked, "Why are you alone?"

阿布从早等到晚，渴了就在水塘里喝几口水，饿了就去附近啃一点快枯的草，困了就在不远处打个小盹，生怕和家人错过。三天过去了，七天过去了，陆续路过的两个马群都不是他的家族，唯一幸运的是，他的天敌也没有出现。他不得不绝望地接受现实——他找不到家人了。

From morning to night, Abu kept waiting near the pond. Seven days past, two groups had passed by, but there were no familiar faces. Fortunately, his enemies also hadn't appeared. Abu had to accept the reality in despair—he couldn't find his family.

天气越来越冷,雪花飘了下来,每年冬天都有不少成年的野马死在卡拉麦里。小马如果失去家族的庇护,几乎不可能撑过寒冷的冬季。阿布蜷缩着身体卧在荒漠里,想象着和家人在一起的快乐时光,"爸爸妈妈,你们在哪里?我好想你们!"

It was getting colder and colder, and snow fell down. Every winter, many adult wild horses died in Kalamaili, let alone a young horse. Abu crouched in the desert, imagining the happy time with his family, "Mom and dad, where are you? I miss you so much!"

孤立无援的阿布乞求别的马群带他一起走,"你们能带我一起走吗?"他问路过的马群,可是没有家族愿意接纳他。水塘结冰了,阿布不得不独自在荒漠里流浪,去寻找新的水源。

"Can you take me with you?" Abu begged other groups helplessly, but no family was willing to accept him. The pond froze, and Abu had to find a new one.

连续走了两天，水源在哪儿呢？可怜的阿布还不到1岁，不像他的家长们那么耐渴，忍受一两天不喝水已经是他的极限了。雪很快就覆盖了这片荒原，更加可怕的是，阿布发现他的身后有一匹狼在尾随，这匹狼好像也落单了，毕竟如果是狼群的话，阿布早就成为盘中餐了。那匹狼不紧不慢地跟着阿布，保持着一定距离，好像在盘算着等阿布体力不支的时候发起攻击。阿布愈发焦躁不安，许久没有喝水和进食让他的体力直线下降。突然，不远的山丘上出现了一匹公马的身影。

After walking for two days, Abu had no gain at all. This poor boy was less than one year old. One or Two days without drinking water was his limit. Snow soon covered the wasteland. What was worse, a wolf was following Abu. Leisurely, the wolf kept a certain distance behind, as if planning to attack when the dish was exhausted. Lack of water and food made Abu weak. Suddenly, a male horse appeared on the hill not far away.

公马看上去很强壮，就像"国王"那样，阿布有些害怕，不敢靠近。但他回头一看，那匹狼已经开始缩短他们之间的距离，似乎看出来阿布体力消耗得厉害。"跟着他，也许能躲过一劫，也许还能找到水源！"对生存的渴望战胜了恐惧，阿布发力朝公马跑去。

The horse seemed very strong, just like King. Abu was a little afraid of him. But when he looked back, the wolf began to shorten their distance! "Follow him, maybe I can escape, and even find water!" The desire for survival surpassed the fear, abruptly Abu ran to the horse.

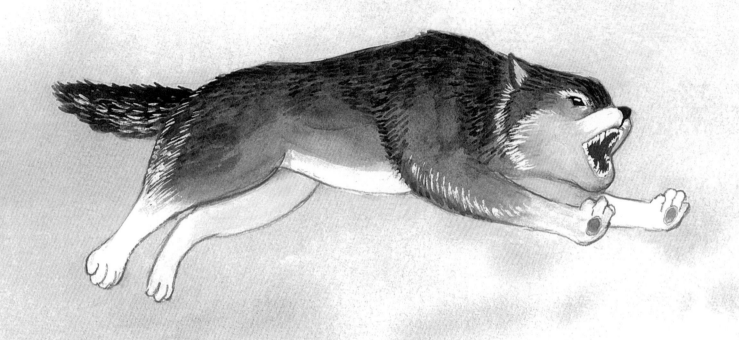

　狼怎么可能放弃马上就要到口的美味？他露出了尖牙，像离弦的箭一样冲向阿布。"救命！帮帮我！"阿布边撒腿奔跑，边高声求助。谁知就在此时，阿布被地上的一块大石头绊了一跤，"咚"的一声摔在地上，扬起了一阵尘土。狼发现有可乘之机，扑向阿布，眼看就要咬上他的脖子。突然！公马卡班挡在了阿布身前，他很早就发现了这两位尾随者。

How was it possible for a wolf to let the feast flee? He rushed to Abu like an arrow. "Help! Help me!" Abu ran and cried. At this very moment, Abu was tripped by a big stone and fell down. The wolf took the chance! When he was about to bite Abu on the neck, Kaban intervened. He had already noticed them.

荒原上落单公马的危险性很大，不得不说阿布的运气很好，他遇到了一匹好心的公马。

It was very dangerous to meet a single stallion on the desert, but Abu was lucky enough—he met a kind one.

卡班静立在阿布前方，耳朵朝着狼的方向，双眼紧紧瞪着狼，打了个响鼻，随后调转方向，抬起后腿，猛烈朝后蹬踢。狼看到体型健硕的卡班，知道自己不是他的对手，夹着尾巴转身跑了。

Kaban stood in front of Abu, with his ears facing the wolf, his eyes staring at him tightly. He snorted, then turned around and kicked with force! The wolf knew he couldn't win, so he went away in dejection.

卡班望向阿布,"小家伙,没事吧?你怎么一个人?"

"谢谢!谢谢您!我叫阿布,我和我的家族走散了。"

回想起当时做的傻事,阿布越想越伤心,"不对,不是走散了,是我偷偷找了地方睡觉,等我睡醒了,他们都不见了。"

豆大的泪珠从眼眶中滚落,眼泪像决堤的河水,不停地往下流,多日的委屈终于得到了宣泄。

Kaban looked at Abu, "Poor child, are you all right? Why are you alone?"

"Thank you! Thank you so much! My name is Abu. I was separated from my family."

Recalling the stupid things he had done, Abu was quite sad, "It was all my fault. I secretly found a place to sleep. When I woke up, they were all gone."

Tears rolled down from his eyes, like the river breaking a dike.

卡班默默守护在阿布身边，他知道这个小家伙吓坏了。等阿布恢复了平静，卡班开口说道："我叫卡班，你先跟在我身边吧。我带你去找水源，也许路上还能遇到你的家族。"阿布充满感激地点点头，心中又燃起了希望。

Kaban stayed with Abu silently. He knew that the little guy was too scared. When Abu calmed down, Kaban said, "My name is Kaban. You can follow me first. I'll take you to the water source. Maybe we'll meet your family on the way." Abu nodded gratefully.

阿布一路跟随着卡班,他对这个好心叔叔的来历很是好奇。

"卡班叔叔,你也是和家族走散的吗?"

"不是,我是成年后主动离开家族的。"卡班微微笑了一下。

"主动离开,为什么啊?"阿布很不解。

"离开是为了更好地生活。小家伙,你以后也会像我一样,所以你要学习的地方还有很多。"卡班耐心地回答道。

Abu was very curious about this kind uncle.
"Uncle Kaban, were you also separated from your family?"
"No, I left my family voluntarily when I grew up." Kaban gave a little smile.
"Voluntarily, why?" Abu was puzzled.
"For a better life. Little boy, you will know, and now you still have a lot to learn." Kaban replied patiently.

阿布跟着卡班一路跋涉寻找着水源和家人。

"水！我们找到水啦！"阿布发现前面有一片水塘，兴奋地跑了过去，咕咚咕咚地大口喝了起来，还不忘礼貌地招呼卡班，"卡班叔叔，你也快来喝啊！"

久违的水源缓解了两匹马长途奔波的劳累。

Abu followed Kaban all the way in search of water and his family.
"Water! There's water!" Abu found a pool and ran over excitedly. "Uncle Kaban, come on!"
The tiredness of two horses after a long-distance running was alleviated.

阿布跟着卡班在被大雪覆盖的荒原上游荡。经验丰富的卡班教会了阿布很多生存技能，其中最重要的就是怎么找到吃的。阿布学会了用前蹄刨开积雪，吃下面的枯草，这可是在冬季的荒原上生存必备的本领。

Abu wandered with Kaban on the snowy moor. Experienced Kaban taught Abu many survival skills. The most important one was how to find food. Now Abu knew how to remove the snow with his forefeet and eat the withered grass below.

他们在荒原上遇到了从北方迁徙过来的蒙古野驴群,双方相安无事。他们也遇到了试图攻击他们的一个狼群,狼群的战斗力和独狼相比,那可是呈指数级的增长。为了保护阿布,卡班和几匹狼缠斗了很久,才终于从狼口脱险。就这样,卡班一直保护着阿布,两匹马儿建立了深厚的感情。

They met a group of Asiatic wild asses migrating from the north. They also met a wolf pack trying to attack them, who were far more dangerous than a single wolf. In order to protect Abu, Kaban fought with them for a long time, and finally got out of danger. Two horses established a deep relationship day by day.

积雪融化是卡拉麦里春天到来的信号，季节性洪水给荒漠里的植物带来了充足的水分，沉寂的荒原开始有了生机。阿布安然度过了冬季，几个月的历练，让他的性子变得沉稳，身形也愈显健硕。他把重回家族这个渺茫的希望藏在了心底，只有当深夜万籁俱寂的时候，他才会仰望天空，回想那个银河高悬的夜晚。

Melting snow was a signal of spring in Kalamaili. Seasonal floods brought vitality back to the ground. After several months of training in winter, Abu became more composed and stronger. He buried the dim hope of returning to his family deep in heart, only when the night was quiet would he look up at the sky and recall the Milky Way of that day.

湛蓝如洗的天空中飘着几朵悠悠的白云，荒原上的野花竞相绽放，就像点缀在一条绒毯上的精美刺绣。阿布在美丽的原野中尽情地奔跑，突然他停住了脚步。他简直不敢相信自己的眼睛，远处的那个马群他很熟悉！他甩甩脖子，背上的鬃毛随着微风轻轻飘扬，那是他的家人！他终于找到他们了！

White clouds floated in the blue sky, and wild flowers were blooming on the desert, just like the exquisite embroidery on a velvet carpet. Abu ran heartily in the beautiful field. All of a sudden, he stopped. Those horses in the distance…he knew them very well! It was his family! He finally found them!

卡班弯下脖子蹭了蹭阿布，微笑着说："去吧，小家伙，好好珍惜和家人在一起的时光，我们有机会再见。"

阿布扭头看看卡班，又看看家人，"卡班叔叔，谢谢您！我永远也不会忘记您的！"阿布依依不舍地和卡班告别，随即朝着家人所在的地方飞奔而去。

Kaban rubbed Abu with his neck. He said with a smile, "Go on, my boy. Cherish the time with your family. We'll see each other again some time in the future."
"Uncle Kaban, thank you! I will never forget you!" Reluctantly Abu bade farewell to Kaban, then galloped towards his family.

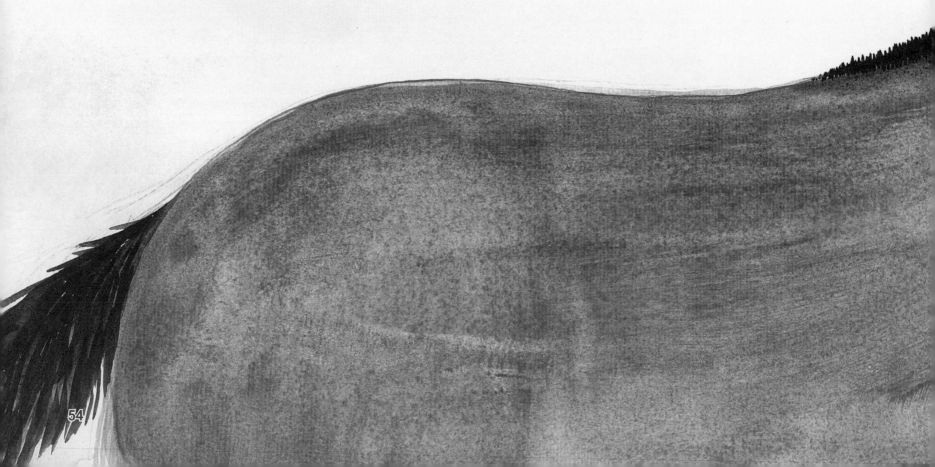

我的家,我的爸爸,我的妈妈,我的兄弟,我回来了!

My home, my dad, my mom, my brothers, I'm back!

野马的秘密知多少

我的名片

普氏野马（学名：*Equus ferus*）隶属哺乳纲奇蹄目马科马属，国家一级保护动物，被《濒危野生动植物种国际贸易公约》（CITES）列入附录Ⅰ物种。普氏野马是目前地球上唯一存活下来的野马，它们的野外种群在我国境内乃至全世界曾经一度灭绝，随着中国政府野马还乡计划的实施，野外种群数量正在逐步恢复中。

为什么叫作普氏野马

1868年,沙俄探险家、地理学家尼科莱·普热瓦尔斯基(Nikolai Przhevalsky)在准噶尔盆地的荒漠上发现了一种土生土长的野马,这让他大为吃惊并一直牵挂于心。他先后3次带领探险队进入新疆,1878年在奇台——巴里坤戈壁上终于捕猎到了第一匹野马,并把标本送回了莫斯科。1881年,沙俄学者波利亚科夫以普热瓦尔斯基的名字给这种野马正式定名为"普氏野马"。在此之前,人们认为世界上真正的野马已经绝迹了,普氏野马的发现在全世界引起了轰动。然而,在被世界关注的同时,普氏野马也受到了更多的威胁,欧美狩猎者们纷至沓来,大规模地捕猎野马。1890年,德国探险家格里格尔从准噶尔盆地捕捉到了52匹野马幼驹,运往德国汉堡,可是长途跋涉让近一半的幼驹半途夭折,最终仅有28匹存活了下来,其中只有不到半数成功繁殖了后代。到20世纪中叶,也就是普氏野马被世人发现的100年后,由于

人类的捕杀和栖息地环境遭到破坏，普氏野马野外种群灭绝。1979年，野马基本处于被圈养的状态，生活在苏联、欧美一些国家的动物园、私人庄园和禁猎区里，数量只有385匹。现在生活在世界各地的普氏野马都是格里格尔当年从准噶尔盆地捕捉到的那些幸存并成功繁殖的12匹野马的后裔。

普氏野马是怎么回到故乡的

普氏野马野外种群的灭绝，引起了国际上对保护该物种的高度重视。经过100多年的圈养和近亲繁殖，导致普氏野马部分基因丧失，并出现了退化的现象。为了保护世界仅存的野马，野马保护管理组织成立了，并制定了两个目标：一是保存现存圈养野马90%以上的遗传多样性；二是将野马重新引入到它们原来生活的环境中。重新引入是最佳的方案，让野马重返原生地这个想法在1978年荷兰阿纳木动物园召开的第一次国际野马会议上被提出。中国和蒙古国因为地域广阔，本身就是野马的原生地，具有野马放归的基本条件，成为实现野马放归野化的两个国家。中国政府承担起了普氏野马种群复壮的使命，在20世纪80年代启动了野马重新引入计划。18匹普氏野马先后从英、美、德等国运回中国，在海外漂泊了100多年的野马的后代终于回到了自己的故乡，回到了祖先生活过的地方。截至目前，全球仅存普氏野马2400多匹，中国拥有600多匹，其中野放种群200多匹。

普氏野马是世界上唯一的野马吗

马属动物起源于距今7500万年以前中生代的爬行动物。它们的原始祖先是6000万年以前新生代的原蹄兽，体型比较矮小，大约只有狐狸那样大。从马的演化过程来看，有始祖马（距今6 000万年的始新世）、渐新马（距今4 000万年的渐新世）、中新马（距今2 500万年的中新世）、上新马（距今200万年的上新世）、现代马（距今2.5万年的全新世）5个演化阶段。随着气候和环境的变化，马的体型变得越来越大，足趾也从5趾变成了单趾，食性和牙齿也都发生了变化。现代马，就是"真马"，推测有欧洲野马、冻原马、森林马和普氏野马4种原始马种。欧洲野马曾广泛分布于欧洲，它们在20世纪80年代末灭绝后，普氏野马成为唯一存世的马的野生种。

普氏野马最大的本领是什么

普氏野马习惯集群生活，通常三五匹或十余匹成群。在冬天的时候，会有更大的集群现象，甚至能达到数十匹，由一匹健壮的公马率领。它们性情凶野，嗅觉和视觉都非常敏锐，甚至在几千米以外发现有什么不对劲，公马就会马上警示，带领马群飞驰而去。野马最厉害的本领在于它们非常耐饥渴，有

时能两三天只喝一次水、吃一次东西，这也是它们为什么能生活在卡拉麦里这样夏季炎热干旱、冬季寒冷积雪极其恶劣的自然环境里的秘密。

普氏野马喜欢吃什么

普氏野马生活在开阔的戈壁荒漠或沙漠地带，以戈壁针茅、驼绒藜、角果藜、假木贼、蒿、猪毛菜、芨芨草等荒漠植物为食。野马并不能任性地随意选择食物，而是要尽可能地利用全部的生存资源。蒙古野驴、鹅喉羚都是它们食物的竞争对手。

普氏野马什么时候战斗力最强

  普氏野马一般由强壮的公马作为首领，首领的地位是通过公平而激烈的竞争产生的。为了争夺头马地位，公马会不遗余力地进行战斗，因而这也是野马战斗力最强的时候。战斗前，争斗的双方会如仇敌相见，怒目而视，如果没有一方退却，便会扑上去互相撕咬，时而你咬我一口，我踹你一脚，时而首尾相接，在荒野上打转，嘴里还会发出粗重的恐吓和示威声。打斗最激烈的时候，两匹野马会忽地立起，前蹄悬空对搏。在一旁观战的母马，最终会接纳战斗的胜利者。失败者会黯然离去，加入"光棍营"等待东山再起的机会。

普氏野马是卡拉麦里最强悍的动物吗

在卡拉麦里,普氏野马的主要天敌是狼。随着当地蒙古野驴和鹅喉羚等有蹄类动物数量的增加,狼的数量也在同步增长。狼是卡拉麦里的顶级捕食者。头马会将3岁左右的亚成年个体驱逐出群,头马更替也会导致公马单独离群,单独离群的野马更容易遭到狼的袭击。此外,卡拉麦里的夏季高温干旱,有限的水源地让野马种群相对集中,这无疑增加了被狼群攻击的概率。

普氏野马和蒙古野驴是好朋友吗

普氏野马与蒙古野驴从动物分类的角度来看都是马科马属动物,有着共同的祖先。在荒漠里,最宝贵的资源就是食物和水。生活在准噶尔盆地的普氏野马和蒙古野驴所需的生存资源接近,活动空间也存在一定的重合。与其说它们是好朋友,不如说是为了在荒漠环境里生存下去,彼此竞争的好兄弟。

蒙古野驴

普氏野马